I0468976

"Get Alarmed, South Carolina"
Lessons Learned from Its Success

Reported by: Carolyn Perroni

This is a Special Report 044 of the Major Fires Investigation Project conducted by TriData Corporation under contract EMW-90-C-3338 to the United States Fire Administration, Federal Emergency Management Agency.

Homeland Security

Department of Homeland Security
United States Fire Administration
National Fire Data Center

U.S. Fire Administration Fire Investigations Program

The U.S. Fire Administration develops reports on selected major fires throughout the country. The fires usually involve multiple deaths or a large loss of property. But the primary criterion for deciding to do a report is whether it will result in significant "lessons learned." In some cases these lessons bring to light new knowledge about fire--the effect of building construction or contents, human behavior in fire, etc. In other cases, the lessons are not new but are serious enough to highlight once again, with yet another fire tragedy report. In some cases, special reports are developed to discuss events, drills, or new technologies which are of interest to the fire service.

The reports are sent to fire magazines and are distributed at National and Regional fire meetings. The International Association of Fire Chiefs assists the USFA in disseminating the findings throughout the fire service. On a continuing basis the reports are available on request from the USFA; announcements of their availability are published widely in fire journals and newsletters.

This body of work provides detailed information on the nature of the fire problem for policymakers who must decide on allocations of resources between fire and other pressing problems, and within the fire service to improve codes and code enforcement, training, public fire education, building technology, and other related areas.

The Fire Administration, which has no regulatory authority, sends an experienced fire investigator into a community after a major incident only after having conferred with the local fire authorities to insure that the assistance and presence of the USFA would be supportive and would in no way interfere with any review of the incident they are themselves conducting. The intent is not to arrive during the event or even immediately after, but rather after the dust settles, so that a complete and objective review of all the important aspects of the incident can be made. Local authorities review the USFA's report while it is in draft. The USFA investigator or team is available to local authorities should they wish to request technical assistance for their own investigation.

For additional copies of this report write to the U.S. Fire Administration, 16825 South Seton Avenue, Emmitsburg, Maryland 21727. The report is available on the Administration's Web site at http://www.usfa.dhs.gov/

U.S. Fire Administration

Mission Statement

As an entity of the Department of Homeland Security, the mission of the USFA is to reduce life and economic losses due to fire and related emergencies, through leadership, advocacy, coordination, and support. We serve the Nation independently, in coordination with other Federal agencies, and in partnership with fire protection and emergency service communities. With a commitment to excellence, we provide public education, training, technology, and data initiatives.

 Homeland Security

ACKNOWLEDGEMENTS

Interviews with the following individuals contributed important information and perspective for this report:

Richard S. Campbell   South Carolina State Fire Marshal

Sondra Vann Senn   Programs Manager, Division of State Fire Marshal, South Carolina Budget and Control Board

Charles Vaughan   Fire Prevention Field Coordinator, "Get Alarmed, South Carolina" Program, South Carolina Division of State Fire Marshal

Mary Lee Maiden   Public Information Manager, South Carolina Division of State Fire Marshal.

ACKNOWLEDGEMENTS

TABLE OF CONTENTS

INTRODUCTION

I n 1988, the State of South Carolina kicked off a public fire safety program entitled "Get Alarmed, South Carolina!" The program has been credited in 1990 with helping the State record the lowest number of fire deaths in five years and to begin the reverse a trend in recent years toward higher and higher fire death rates. The story of the "Get Alarmed, South Carolina!" program and other South Carolina public fire safety efforts since 1988 offer lessons which may be helpful to other States.

BACKGROUND

In 1988, fire claimed the lives of 164 people in South Carolina. With just over 47 fire deaths per million population, South Carolina had the highest fire death rate in the country. Prior to 1988, South Carolina had consistently ranked among the top three States with the highest fire death rates in the Nation. In addition, the rate had been escalating at an average of 20 percent per year.

Data on fire deaths collected by the South Carolina Budget and Control Board's Division of State Fire Marshal indicated that a disproportionate share of these fire deaths occurred in the State's rural and economically depressed areas. As in many parts of the country, the high-fire-risk groups in South Carolina were the poor, the elderly, and the handicapped.

State Fire Marshal Richard S. Campbell and his staff began in 1987 to plan for a major fire prevention program aimed at reversing the fire death trend. A smoke detector giveaway was chosen as the centerpiece for the program. There was statistical and anecdotal evidence nationwide to show that properly installed and maintained smoke detectors saved lives by providing the early warning necessary to allow people to escape deadly fires. In addition, South Carolina's data showed that many of the fatal fires in the State occurred in homes without detectors and that many of the fire deaths occurred at night as people slept – the time when smoke detectors are most helpful.

The Division of State Fire Marshal carried out a small-scale program in 1986-87 that indicated a smoke detector giveaway program had the potential for success. The Division had re-programmed about 9,500 dollars to supply 20 fire departments across the State with 45 smoke detectors each. The detectors were offered to fire departments in rural or economically depressed areas. The departments were asked to join forces with some community organizations and seek local funds for the purchase of additional smoke detectors; to distribute detectors to the poor, elderly, and handicapped in the community on a priority basis; and to issue annual reminders to smoke detector recipients about the importance of detector maintenance.

This program lasted two years. During that time, many local fire departments joined with community volunteer organizations to obtain funds through a Community Volunteer Fire Prevention Program, funded by the United States Fire Administration (USFA) and administered by the National Criminal Justice Association.

Nearly 13,000 additional smoke detectors were purchased over the two years, building a program that delivered about 115,000 dollars worth of smoke detectors into people's homes out of an original investment at the State level of 9,500 dollars.

"GET ALARMED!"

The success of the small scale program convinced the State to proceed with a larger effort. The "Get Alarmed, South Carolina!" program began in 1988.

The core of the program was a smoke detector giveaway, primarily in high-risk areas. To support the giveaway, a massive effort was undertaken Statewide to build public awareness about the fire problem and educate the public about potential fire hazards.

To give further emphasis to the program, South Carolina Governor Carroll Campbell proclaimed 1988 as "Fire Safety Awareness Year" in the State. First Lady Iris Campbell was named as Grand Marshal for the "Get Alarmed" campaign.

South Carolina's position among the top three States with the highest fire death rates and the success of the earlier smoke detector giveaway effort helped convince the State legislature to appropriate 50,000 dollars to fund the project.

SMOKE DETECTORS

About half the appropriated funds went to purchase smoke detectors. The Division of State Fire Marshal bought the detectors at a State contract price of about 4.50 dollars each.

The smoke detectors served as an incentive for the fire service in the State to get involved. One hundred detectors were offered free to each county in the State (200 were offered to each of the four most populous counties) under the same conditions as had been used in the 1986-87 pilot program. They had to agree to:

- join forces with some community organization to promote the program and leverage funds for the purchase of additional smoke detectors;

- give priority to distributing to the poor, elderly, and handicapped in the community; and,

- issue annual reminders to smoke detector recipients to reinforce the importance of detector maintenance.

As added support, any local government participating in the "Get Alarmed" program was allowed to buy additional detectors at the low State contract price.

FIRE PREVENTION MATERIALS

The other half of the funds appropriated for the "Get Alarmed" program, plus additional funds from the Division of State Fire Marshal budget, were used to produce a variety of fire prevention materials. This included posters, a calendar with a separate fire safety message for each month of the year, and flyers for each month. In some cases, the fire safety messages and artwork to illustrate them were created specifically for the "Get Alarmed" program; in other cases, they were adapted from existing materials, such as the fire prevention awareness campaign kits available from the USFA. The materials were supplied free of charge for distribution by local governments, fire departments, and other organizations participating in the program. (See Appendix A.)

In addition, "Get Alarmed" fire safety messages were carried on about 200 billboards throughout the State. The billboards were donated by their owners or paid for by other organizations; the State only had to pay for the printing.

Six video public service announcements (PSAs) were produced and distributed to all major television stations serving the South Carolina market. The PSAs featured an introduction by the First Lady and a short orientation to the "Get Alarmed" program, as well as specific fire safety messages. Topics for these safety messages included installing smoke detectors, testing and maintaining detectors, developing and planning exit drills, and preventing heating and cooking fires.

A package of 22 radio PSAs also was produced and sent to every radio station in the State. As in the case of the printed materials, some of the PSAs were created specifically for the "Get Alarmed" program, and some were adapted from existing material supplied by the USFA and others.

COUNTY COORDINATORS

Grand Marshal Campbell appointed local community leaders, many of whom had fire service backgrounds, as county coordinators for the "Get Alarmed" program. Their appointments often served as media events, with the First Lady formalizing the appointment in ceremonies in the county or at the Capitol. This not only provided a tailor-made "photo opportunity" through which to publicize the campaign, but also offered a way to involve and reinforce community and fire service leaders in the program.

County coordinators were given three main responsibilities:

- contacting each fire department in the county and encouraging them to participate in the "Get Alarmed" campaign;

- distributing smoke detectors and fire prevention literature to participating departments; and,

- seeking funds for purchase of additional smoke detectors.

The success of their efforts was a major boost to the program. Local citizens, organization leaders, and community officials seemed to be more receptive to the program because of the endorsement by and involvement of someone from their own area.

MEDIA SUPPORT

Media support became a very important element in the "Get Alarmed" program. News releases and articles about the program were sent out routinely by the Division of State Fire Marshal to media contacts throughout the State. For this program, an additional copy of the material was sent to participating fire departments, so they could make local contact with media representatives and reinforce the importance of the program and provide additional information about local program activities.

The First Lady's involvement as Grand Marshal provided additional opportunities to draw the media's attention to the program. Her attendance at "Get Alarmed" meetings and visits to individual communities to "kick off" the program locally helped build media interest throughout the State.

Joe Pinner, a well-known radio and television personality in the State, gave the program an additional boost. Pinner not only talked about smoke detectors frequently on his radio and television shows, but also wrote a letter about the "Get Alarmed" program and encouraged his colleagues throughout the State to use the PSAs produced for the program. The Division of State Fire Marshal sent his letter along with the PSAs when they were mailed to radio and television stations, and his endorsement helped get the PSAs used in many cases.

An annual "birthday party" for the "Get Alarmed, South Carolina" program is a media event in the State. It not only serves to focus public attention on the need to clean detectors and change batteries, but also helps renew interest and involvement in the program. Fire departments use the event to kick off local publicity campaigns on smoke detector maintenance.

PARALLEL EFFORTS

While the "Get Alarmed, South Carolina" program was being planned and initiated, other fire prevention efforts were in process. The synergy of these efforts has helped the State succeed in reducing fire deaths.

In 1988, a major strategy development conference was sponsored by the South Carolina State Fire Commission and the Division of State Fire Marshal assisted by funding from the USFA. The conference drew some 80 participants who represented 24 State agencies, the medical profession, insurance companies, public utilities, private industry, the fire service, civic and service organizations, the Federal Emergency Management Agency (FEMA), and National fire service organizations. The purpose of the meeting was to discuss South Carolina's fire problem in the context of three main topic areas: public fire safety education; use and misuse of portable and solid-fuel heating appliances; and responsibilities toward children, the elderly, and the handicapped.

The conference attendees developed consensus recommendations in all three topic areas and endorsed the formulation of a Fire and Life Safety Task Force to design a strategy for implementing the recommendations and achieving a lower fire death rate in the State. The conference among other things spawned efforts by several State agencies, including the Commission on Aging and the State Department of Agriculture, to regularly coordinate with the Division of State Fire Marshal and incorporate fire safety messages into programs for their respective constituents.

This coordination extended and reinforced efforts to reach the public through the "Get Alarmed" program. Specifically, coordination between health and social services agencies and the Division of State Fire Marshal made it possible to reach people in high-risk groups through agencies and programs that have routine contact with individuals in these groups. Personnel from agencies that provide in-home care – the Department of Social Services, Department of Health and Environmental Care, and Commission on Aging, for example – were schooled in the "Get Alarmed" program concept and provided with fire prevention materials to distribute as they made their usual rounds.

The "Get Alarmed" program and its fire safety messages also were integrated into the State's "Teach the Teacher" program. This is a yearly seminar targeted at fire safety educators at the local level. The "Teach the Teacher" seminar program had begun in 1986-87 as a way to showcase local programs developed under the Community Volunteer Fire Prevention Program and give fire safety educators from other communities the information and motivation they needed to replicate these efforts. It served a similar purpose for the "Get Alarmed" program and helped reinforce the necessity for building local cooperation to prevent fire deaths.

PROGRAM RESULTS

The results achieved in the first two full years of the "Get Alarmed, South Carolina" program are eye opening. More than 25,000 smoke detectors – five times the number originally funded by the State – were placed in homes throughout the State, and more than two-thirds of the State's 700 fire departments got involved in smoke detector installation and maintenance programs in their communities. In addition, there is a broader awareness among the citizens of South Carolina about fire safety and the value of smoke detectors.

The increased awareness generated by the "Get Alarmed" program undoubtedly was a major factor in the marked drop in actual fire deaths that South Carolina experienced in 1990. Moreover, as shown in the figure below, the "Get Alarmed" program can be credited with saving many lives that might have been lost if the program had not been undertaken, since it successfully interrupted a record of 20-percent-per-year increases in the State's fire death rate.

ESTIMATED LIVES SAVED SINCE
INITIATION OF "GET ALARMED, SOUTH CAROLINA"

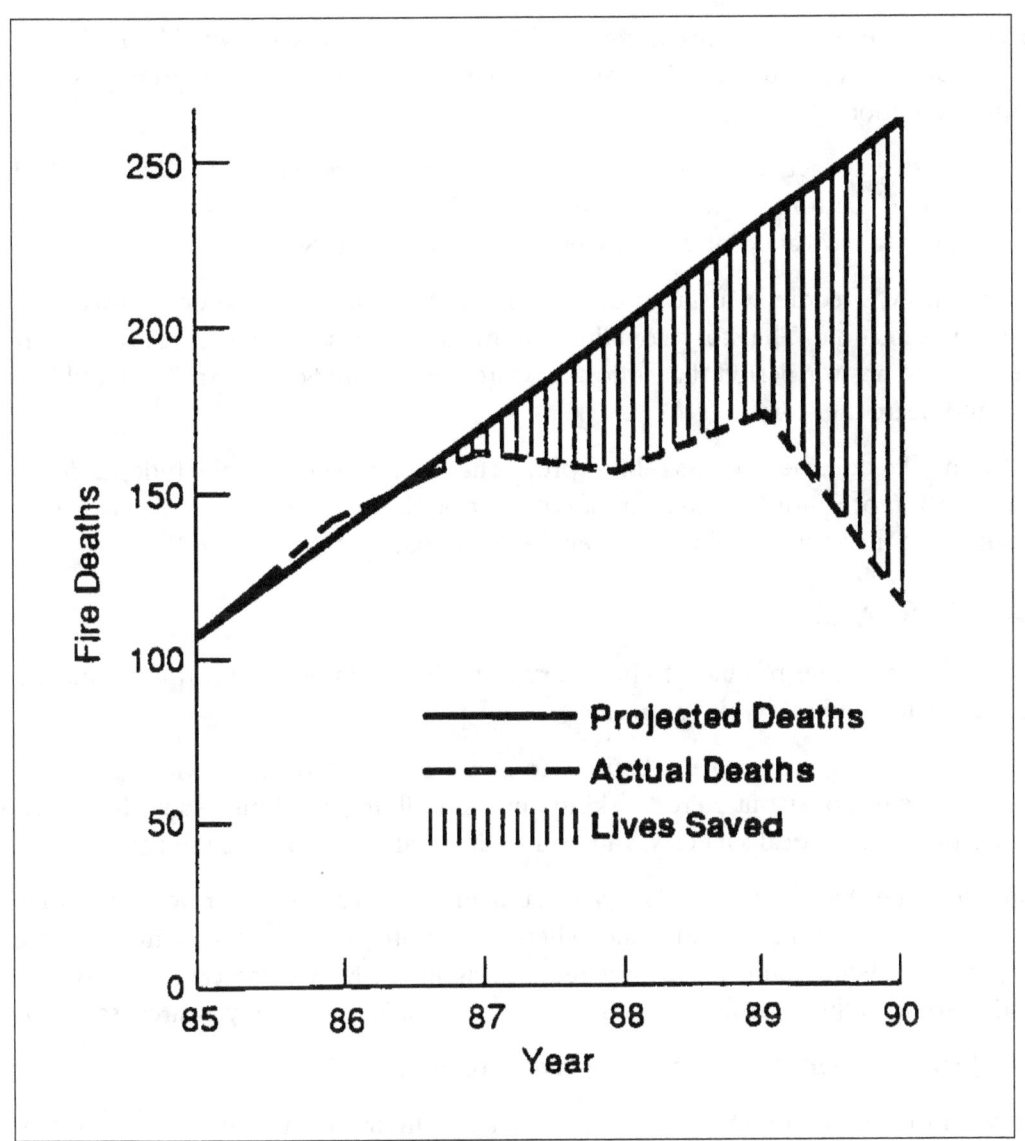

Furthermore, the program continues to be viable. Fire departments and community leaders appear to have built relationships that will work for years to come. The cooperation among local organizations continues to make it possible for fire departments in the State to make more use of fire prevention programs and materials offered by the State.

In fact, the State has launched a new follow-up to the "Get Alarmed" program – this time without any additional appropriation. The new program is called "Save Our Children..Get Alarmed, South Carolina." It's objective is to get hospitals to include information about smoke detectors in pre-natal education classes and to give smoke detectors as gifts to parents of new babies. The program is being totally funded by local communities, in most cases using the fundraising alliances they built as part of the original "Get Alarmed" program. For example, participants in this program include the Beaufort County Memorial Hospital and the Beaufort Fire Department, the Wallace Thompson Hospital and the Union County Firemen's Association, and the Newberry Memorial Hospital and the Newberry County Firemen's Association.

Another effort to prevent fire deaths among children involves a mascot called "Freddie the Fire-less Feline." The Division of State Fire Marshal introduced "Freddie" to the children of South Carolina during Fire Prevention Week in 1989.

"Freddie" is a big, gray cat that wears yellow boots, a red firefighter's hat, and a black and white cloth coat. The "Freddie" costume is worn by Division of State Fire Marshal staff members, firefighters, and others who visit day care centers and schools throughout the State.

A story about how "Freddie" was saved by a firefighter in a house fire when he was a kitten is the core of the presentation. The story contains a number of fire safety messages that are reinforced by a question and answer period that follows. More than 5,000 South Carolina children have met "Freddie" and heard the story.

Funding for the "Freddie the Fire-less Feline" project has been provided by the Independent Insurance Agents Foundation of South Carolina, and a coloring book featuring "Freddie" has been produced with a grant from the South Carolina Firemen's Association.

LESSONS LEARNED

1. **Limited funding does not have to be a barrier to development and implementation of good fire prevention and safety programs.**

    If there is one thing the people of South Carolina demonstrated in the "Get Alarmed" program, it is that there is money out there. Seeking out and building relationships with civic and service organizations, other local agencies, and even individual benefactors is the key.

    In addition, the Division of State Fire Marshal made good use of fire prevention resources available free of charge from the USFA and others. For example, the expense involved in designing new public fire education materials was reduced significantly by using camera-ready artwork and adapting some of the fire safety materials provided in USFA's fire safety awareness campaign kits.

2. **Paid television ads may be as cost-effective as television PSAs.**

    Even with the endorsement of a well-known media figure in the State, it proved very difficult to get good air time for the PSAs developed for "Get Alarmed, South Carolina." This is not a problem unique to this program; PSAs, particularly those offered by local and State governments often are relegated to late-night or early-morning, non-prime viewing hours. The impact of the ads is diluted because of the lack of audience. Since quality television ads are expensive to produce and distribute, it may be more reasonable to produce one good ad and pay for a few good time slots on local or Statewide television that would offer the potential for reaching your intended audience.

3. **Having a well-known spokesperson helps the profile of the program.**

Having South Carolina's First Lady as the Grand Marshal for the "Get Alarmed" program proved to be helpful in garnering media attention for program activities and lent credibility to the program overall. In addition, her participation appeared to be pivotal in leveraging donations and in-kind contributions from businesses Statewide.

4. **The strong, multi-year commitment of the State Fire Marshal helps the program succeed.**

The sustained commitment of State Fire Marshal Richard S. Campbell and his staff was a major factor in the success of the "Get Alarmed, South Carolina" program. Beginning in 1987, the Division of State Fire Marshal made reducing the State's fire death rate by increasing smoke detector usage a consistent priority. This commitment was demonstrated not only through appearances by the State Fire Marshal himself at local "Get Alarmed" program events throughout the State, but also through the Division of State Fire Marshal's continuous day-to-day support for the county coordinators for the "Get Alarmed" program.

The division also carried out an aggressive campaign to get the media to focus more attention and devote more coverage to the fire death problem in the State. They encouraged fire departments to reinforce State-level efforts by making contact with their local media representatives. In addition to benefiting the "Get Alarmed" program, this has helped local fire departments build a cooperative relationship with the media that can assist in future fire prevention efforts.

The Division of State Fire Marshal made ready-to-use "Get Alarmed" program materials available at no cost to local fire departments. This enabled many fire departments to participate in the program – departments which would not have had the resources to produce materials on their own.

APPENDIX A

Following is a selection of public education materials and PSAs produced for the "Get Alarmed, South Carolina" program. Some were written specifically for the "Get Alarmed" program; others were adapted from existing public fire safety education resources. You are encouraged to use these items, or adapt them as necessary, to expand your own State or local smoke detector promotion or other fire safety awareness programs.

Materials

1. "What You Need To Know About – Smoke Detectors, Heating Safety, Cooking Fires, Cigarette Smoking, Escaping a Fire"
2. "Give the Gift of Life...Give a Smoke Detector"
3. "Smoke Detectors are Life Protectors"
4. "Get Alarmed, South Carolina! A Fire Safety Awareness Campaign"
5. "Smoke Detectors Save Lives...They Could Save Yours!"
6. "Holiday Safety"
7. "Plan Your Escape"
8. "Stay Warm Safely"
9. "Their Life Is Your Responsibility"
10. "School's Out!"
11. "Check Your Hot Spots"
12. "Put A Lid On Cooking Fires"
13. "Don't Let A Fire Take The Bang Out Of Summer"
14. "Freddie the Fire-less Feline"

PSAs

1. "Summer Fun"
2. "Smoke Detectors"
3. "Smoke Detectors Message For Parents"
4. "Smoke Detector Installation"
5. "Smoke Detector Maintenance"
6. "Get Alarmed Campaign"
7. "Staying Alive"
8. "Survival"
9. "After School Kids"
10. "Fire and Drinking"
11. "Cigarette Smoking Fires"
12. "Holiday Safety"
13. "Fire Drills"
14. "Stop, Drop, and Roll"
15. "Handle With Care"
16. "Smoke Kills"
17. "Cooking Fires"
18. "Steaks Are For Cooking"

GET ALARMED

SOUTH CAROLINA!

WHAT YOU NEED TO KNOW ABOUT —

- Smoke Detectors
- Heating Safety
- Cooking Fires
- Cigarette Smoking
- Escaping a Fire

Division of State Fire Marshal
Budget and Control Board
1201 Main Street, Suite 810
Columbia, SC 29201
Telephone: 737-0660

A MESSAGE FROM THE STATE FIRE MARSHAL

The Division of State Fire Marshal of the Budget and Control Board, through its fire prevention and fire safety awareness programs, fulfills a legislated responsibility for the prevention of fires and protection of life and property in South Carolina. However, success in reducing fire-related deaths, injuries, and property losses can only come when individuals become educated about the fire problem in their state and begin to practice fire safety in their homes, schools and businesses. The Division of State Fire Marshal, along with the fire service, is dedicated to making South Carolina a fire safe state. We challenge each citizen to learn more about fire, practice fire prevention, install smoke detectors, and, if a fire should occur, know how to escape. The information in this pamphlet could save your life.

R.S. Campbell, P.E.
State Fire Marshal

SMOKE DETECTORS-LIFE PROTECTORS

Today, we are told 75 percent of the homes in the United States have smoke detectors. Even though they have been around only about ten years, two-thirds of the homes nationally have at least one. In South Carolina, the percentage of detector-protected homes is less. In most communities, new homes are required to have smoke detectors. In South Carolina, State law requires that owners install smoke detectors in the living unit of each apartment building if the complex has more than two apartments.

Why all the interest in smoke detectors? They save lives -- easily and inexpensively. When a fire occurs, chances for survival are two times better in a home where smoke detectors are present.

The plain terrible fact is that smoke kills. Smoke from a fire several rooms away -- smoke in the middle of the night while you're asleep -- smoke loaded with poison gases from burning materials -- smoke that sneaks around your home, slipping around corners and around closed doors -- smoke, that in spite of what you might believe, will not wake you up! It will kill you first.

Most fire victims never see, hear, or feel the fire. They are suffocated by toxic gases long before the fire gets to them. In fact, even after firefighters find the bodies of victims, the people are usually not burned.

So, that's why you need smoke detectors. They stand guard over your family and home. If a fire starts, the smoke detector smells it first and sounds a loud alarm so you can escape in time. There are even some designed for the hearing impaired.

If your smoke detector sounds annoying false alarms; for example, when you broil meats in your kitchen, don't pull out the battery, or disconnect the detector. Simply moving the detector to a new location may prevent the false alarms and better protect your home.

You do have smoke detectors, don't you? You should have at least one, located between your sleeping areas and living areas. That's where you are most likely to be when the fire starts, and that's away from the rest of the house where the fire usually starts. If your bedrooms are in separate areas, each should have its own detector protecting it from living areas. Smokers should have detectors right in their bedrooms, just in case. Pathways to basements, garages and other storage areas should have a smoke detector standing guard.

Are you unsure about the different types of detectors or need more specific information about where the detectors should be placed? Read on.

EXCERPTS FROM:

"Smoke Signals" - Second Edition
U. S. Consumer Product Safety Commission
Federal Emergency Management Agency

HOW DO THEY WORK?

There are two basic kinds of smoke detectors -- ionization and photoelectric. In the ionization detectors, a small and carefully shielded bit of radioactive material "ionizes" the air in the detector's smoke chamber. As a result, a very weak electrical current flows through the chamber and is sensed by the detector's circuit. When tiny particles of smoke drift into the chamber, they reduce the electrical flow. When enough particles have entered the chamber, the electrical current drops below the acceptable threshold and the detector circuit turns on the alarm buzzer.

The other most frequently purchased type of home smoke detector uses the photoelectric principle. It detects smoke by "seeing it" in much the same way your eyes do -- by means of light reflected by the particles of smoke. Some of the reflected light falls on a photocell causing it to produce a slight electrical current. When the smoke particles are dense enough to reflect a pre-set amount of light, the detector circuit actuates the alarm.

Generally speaking, since hot blazing fires tend to produce smaller smoke particles which float in rising hot air from a fire, ionization detectors usually have a slight edge in giving early warning of open, flaming fires. Smoldering fires produce large particles so photoelectric detectors are somewhat more likely to give the alarm while a fire is smoldering. But remember, many household fires produce detectable amounts of both visible and invisible smoke. Either kind of detector has a high probability of giving you enough warning for safe escape.

PLACE ONE DETECTOR ON EVERY FLOOR

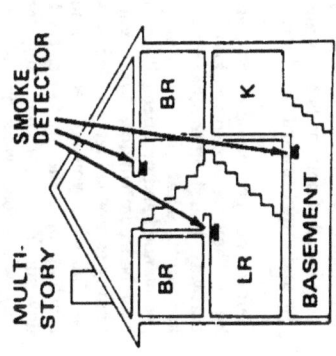

MULTI-STORY

SMOKE DETECTOR

BR · BR · LR · K · BASEMENT

SINGLE LEVEL

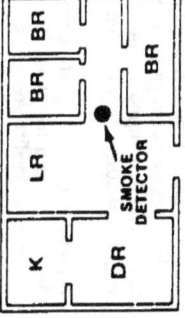

K · LR · BR · BR · DR · SMOKE DETECTOR · BR

Make sure detectors are placed either on the ceiling or 6-12 inches below the ceiling on the wall. Locate smoke detectors away from air vents or registers; high air flow or "dead" spots are to be avoided.

WHICH IS BETTER -- ELECTRIC OR BATTERY-POWERED?

Battery operated detectors are less expensive, easier to install. The detector needs dusting with a vacuum cleaner often to remove dust and cobwebs. You need to test the detector monthly to be sure it is operable. Batteries need to be replaced every year. In about a year after installation, a detector begins to emit "beeps" every minute or so, and will keep this up for a week or longer to tell the owner the battery power is low and should be replaced.

Detectors which operate on household electric current have the power they need as long as there is current in the circuit to which they are connected. Installation is somewhat more complicated because sometimes the ideal location for the detector is not near a convenient outlet. In these instances, there will be the expense of having an electrician install an outlet. In the event of power failure, electric detectors will become inoperable (except for some brands, which contain stand-by batteries). It is also possible that a fire could actually start in a circuit which supplies power to the detector, and the power to the unit might fail before it gave an alarm.

HOW MANY?

Tests conducted by the National Bureau of Standards have shown that two detectors, on different levels of a two-story house are twice as likely to provide an adequate amount of time for escape as one detector. The upstairs detector senses smoke wherever it originates, while the downstairs unit will react sooner to fire which could block escape route through the first floor. For minimum protection, install a smoke detector outside of each bedroom or sleeping area in your home.

WHERE DO YOU PUT IT?

Instructions are included on the box and should be followed carefully when installing the detector. Most will recommend installing the detector on the ceiling or on walls between six and 12 inches below the ceiling. There are several places NOT to put the detector. Don't put it within six inches of where the wall and ceiling meet on either surface. This is a dead air space that gets little air circulation. Do not install in front of an air supply duct outlet or return. Also avoid putting detectors on ceilings which are substantially

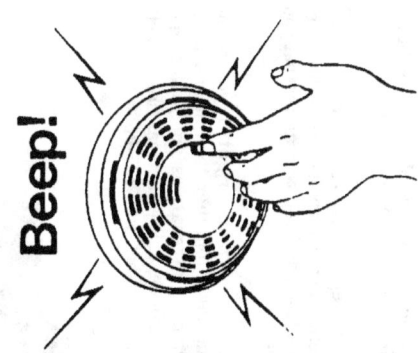

Beep!

warmer or colder than the rest of the room. An invisible thermal barrier near the surface could prevent smoke from reaching the detector. This can be a problem in mobile homes or older, poorly insulated homes.

REMEMBER

Now, you know all about smoke detectors. Install it. Test it monthly. Clean it from time to time. Keep batteries or photoelectric lamps on hand to replace as needed. Your smoke detector is one of the most significant improvements in home safety ever to occur.

You need to help your smoke detector save your life by learning how to escape from fire. Make a family escape plan together and practice it often. □

KNOW HOW TO GET OUT AND DON'T COME BACK!

You're always careful about fire safety. You never smoke in bed. You are cautious about electrical and heating appliances. You're careful when cooking. You have smoke detectors, test them every month, and replace batteries every year. You never expect to have a fire in your home. But just suppose, one night very late when your family is sleeping, a fire starts in your home. Consider what might happen.

A small fire has started. It smolders for awhile, but soon your smoke detectors begin to sound. You wake up to loud, harsh alarm. As you awaken you realize you can smell smoke and hear faint cracking sounds.

Keeping low, you roll out of bed into a crouching position on the floor. You remember reading that air in the upper half or two-thirds of the room is full of poisonous gases and other smoke products. So you stay low and crawl toward the door.

Before you open it you reach your hand up and feel the door and the knob. They are hot so you know the fire is close by and you can't go that way. You shout to other members of the family to alert them. Each responds that he is doing o.k. and heading for the pre-planned and practiced alternate exits. You turn and crawl toward your bedroom window. When you get to the window you pull up the folding escape ladder you bought at the neighborhood hardware store last fall when you noticed how surprisingly inexpensive it was. You reach up, open the window as the top end remains firmly over the window sill. You know your daughter will get out of her room the same way, while your two sons will crawl out their window onto the porch roof and climb down by themselves as they have practiced.

You crawl out of your window and down the ladder. You head straight to the end of your driveway where you all have planned to meet in case of fire. Your sons are running up as you approach the spot. Your

teenage daughter is walking toward you. She volunteers to go next door to call the fire department. Your youngest son remembers his pet lizard and is disappointed, but he knows he can't go back inside to get it.

Your daughter is back in minutes and reports that call has been made. Soon you hear the sound of the siren. The trucks pull up to your house. You can see the fire through the windows, and you're grateful that you and your family are safely outside.

That's exactly how it would happen at your house, right? If you're not sure it would go that smoothly, sit down together tonight and work on an emergency escape plan, including alternate escape routes, and a meeting place. If you need some advice, call your local fire department, or the Division of State Fire Marshal, 803-737-0660. ( )

COOKING FIRES

Many of the deaths and injuries resulting from cooking fires could have been prevented if people were more careful and if they knew how to react to a cooking fire.

Most cooking fires start with grease. Cooking oil and melted fats from meats are flammable when they get so hot they flare up. Be extra careful when cooking with oil or juicy meats. Keep the heat as low as possible. AND NEVER LEAVE COOKING UNATTENDED, EVEN FOR A MINUTE.

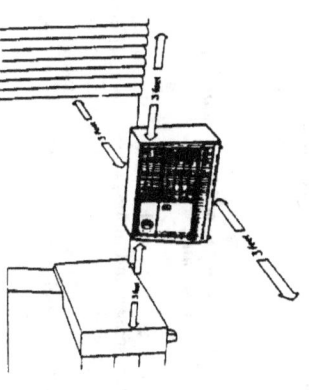

If grease ignites while you're cooking, you must smother the fire quickly and carefully. If the fire is in a pan on the stovetop, smother it by sliding a lid or large pan onto the top of the fire.

NEVER throw water or flour on a flaming pan. Water will spread the flames and flour will actually explode. Do not try to pick up the pan to carry it outside. Do not turn on the exhaust fan over the stove. If the fire is in the oven, turn off the heat and leave the door closed to cut off the fire's air supply.

You can use a portable fire extinguisher on a cooking fire. A fire extinguisher is an excellent safety appliance for the kitchen. But don't wait until you have a fire to learn how to use it. There won't be time to read the directions!

If the fire is too large to safely cover with a lid or for your fire department extinguisher, leave the house immediately and call the fire department from a neighbor's phone. Trying to fight a fire yourself when it's too large can cause far more extensive damage and result in serious injury.

Practice good fire prevention by keeping your cooking appliances clean. Grease build-up can flare up easily. Keep combustibles, paper towels, dish towels, paper bags, away from the stove or other heat source.

Do not wear long flowing sleeves while cooking. The clothing could easily touch a burner and ignite.

Always turn pot handles inward to avoid spills.

Keep children away from stoves.

It's always a good idea to check stoves and other appliances before going to bed or leaving your home to make sure that everything has been turned off. □

HEATING
BE CAREFUL HOW YOU HEAT

Statistics compiled by the Division of State Fire Marshal point to heating as the number one cause of fire-related deaths in South Carolina. Many families use portable heaters to supplement their home heating systems. In some instances, families are forced, because of economic conditions, to try to heat their home with only portable heaters. Some use wood stoves incorrectly; some fail to have chimneys cleaned. Many do not have the basic knowledge about fire safety to protect themselves. It is important for everyone to understand the following:

Keep all heaters at least 3 feet away from anything that will burn, such as curtains, furniture, bedding, paper. Also, keep electric heaters away from sinks, tubs, showers and other containers of water.

Use only a heater that has a safety device to switch it off automatically if it tips over. Make sure that the heater has a UL listed tag showing that it has passed a safety test.

Check the cord of electric heaters before plugging it in. If there is any sign of frayed cords or worn or broken spots, have the cord replaced. Do not simply tape over the worn spot -- it is not enough to prevent a fire. Avoid using extension cords with portable heaters.

Keep children and pets away from portable heaters. Do not place a heater anywhere it could block your escape in case of fire.

Use kerosene heaters according to manufacturer's instructions. Be sure there is a window slightly open to provide ventilation. Never use anything but 1K kerosene which is clear like water. Do not fill the tank inside your home. Spills can mean disaster.

Have a professional service your heating system every fall.

THE SURGEON GENERAL ISN'T THE ONLY ONE

The U.S. Surgeon General says smoking is hazardous to your health. The State Fire Marshal thinks so, too -- in a different way.

One of the most common causes of fire deaths in the home is careless smoking. Nationally, about one-third of all home fire deaths and one-fifth of injuries start with cigarettes. These fires cause more than $3 million in property losses.

If you have smokers in your home, use the following precautions: Provide large, deep ashtrays for smokers. Don't balance ashtrays on arms of furniture or other narrow spots. Be sure ashes are completely cool before emptying ashtrays and then never into trash cans.

Most smoking fires are started when a hot cigarette, ash, or match drops onto upholstery, bedding, carpet, or clothing. There it can smolder for more than a half hour before even a tiny flame starts. All the while the burning cigarette and fabrics give off smoke and deadly fumes.

Never smoke in bed or while reclining in comfortable upholstered furniture. It's too easy to fall asleep. Check all seat cushions before going to bed.

Install extra smoke detectors in smokers' bedrooms and other rooms where they spend a lot of time.

The combination of smoking and drinking alcohol is doubly deadly. Drinking dulls the senses and induces deep sleep. If you smoke while under the influence, you may never wake up and get out if a fire should break out. A house that has just hosted a party is high risk. A cigarette dropped into cushions could smolder for hours, waiting for everyone to be asleep, before erupting into flames. []

11

Have your chimney checked and if needed, cleaned by a professional every year. A substance called creosote builds up in the chimney over time and can start a chimney fire. Older chimneys should be inspected for cracks and flaws and corrected.

Woodburning stoves should be installed by a professional according to manufacturer's instructions. Know that a loud roar, sucking sounds, or shaking pipes means that your wood stove is overheated or there is a chimney fire. Quickly shut off the fire's air supply by closing any air intake vents in the firebox. Close the damper. Get out, and call the fire department from a nearby phone.

Never use a flammable liquid, such as gasoline, to start a fire or put on a dying fire to rekindle.

Keep a tight-fitting screen or glass doors in front of your fireplace at all times. Do not use charcoal in your fireplace.

Dispose of ashes in metal containers, never paper bags, cardboard boxes or plastic wastebaskets. Wet ashes to cool thoroughly and take outside away from house, garage, carport. Remember, ashes can retain enough heat for several days to cause a fire. Take no chances. []

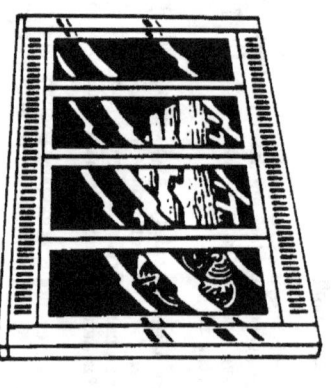

GLASS FIRE SCREEN

10

Give the Gift
of Life . . .

Give a Smoke Detector . . .

The Night
Before Christmas

T'was the
night before
Christmas, when
all through the
house ● not a
creature was stirring,
not even a mouse. ●
When down through the
chimney, all covered with
soot ● Came the "Spirit of
Fire," an ugly galoot. ●
His eyes glowed like embers,
his features were stern ● As
he looked all around him for
something to burn ● What he
saw made him grumble — his anger
grew higher ● For there wasn't a
thing that would start a good fire. ●
No door had been blocked by the big
Christmas tree ● It stood in the corner,
leaving passageways free. ● The lights
that glow brightly for Betty and Tim ●
Had been hung with precaution, so none
touched a limb ● All wiring was new, not
a break could be seen ● and wet sand at its
base kept the tree nice and green. ● The tree
had been trimmed by a mother insistent ●
That the ornaments used should be fire resistant
● the mothers had known the things to avoid
● Like cotton and paper and celluloid. ●Rock
wool, metal icicles and trinkets of glass ● Gave
life to the tree — it really had class. ● And,
would you believe it, right next to the tree ●
Was a suitable box for holding debris — ●
A place to hold wrappings of paper and string.
● From all of the gifts that Santa might bring. ●
The ugly galoot was so mad he could bust ●
As he climbed up the chimney in utter disgust!
● For the folks in this home had paid close
attention ● To all
of the rules
of GOOD FIRE
PREVENTION.

Brought to you through the "Get Alarmed, South
Carolina!" campaign.
Budget and Control Board
Division of State Fire Marshal
1201 Main Street, Suite 810
Columbia, S.C. 29201
803-737-0660

**GRAND MARSHAL
IRIS CAMPBELL**

Get Alarmed, South Carolina!

A Fire Safety Awareness Campaign

WHAT IS THE GET ALARMED, SOUTH CAROLINA! PROGRAM?

The Get Alarmed program is a fire safety awareness campaign with its main objective to lower fire deaths in South Carolina. There are two main goals associated with the campaign: (1) to alarm citizens about the fire problem and teach about fire safety through a public awareness campaign, and (2) to provide smoke detectors to fire high risk citizens.

WHY IS THIS PROGRAM NEEDED?

For several years, South Carolina has had a high national ranking in fire-related deaths. For the last three years, South Carolina has been in the top three states in the nation in fire-related deaths. With 168 fire deaths occurring in 1987, the State ranking went to number one. A program to provide immediate relief to the fire problem was needed. To reach a short range goal, smoke detectors could be the answer. It is a fact that smoke detectors save lives. A person's chance of survival 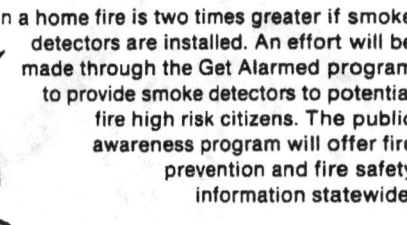 in a home fire is two times greater if smoke detectors are installed. An effort will be made through the Get Alarmed program to provide smoke detectors to potential fire high risk citizens. The public awareness program will offer fire prevention and fire safety information statewide.

WHEN WILL THIS PROGRAM HAPPEN?

Governor Carroll Campbell proclaimed 1988 as "Fire Safety Awareness Year" in South Carolina. He said, "Let's make 1988 the year South Carolina got alarmed and did something about the fire problem." First Lady Iris Campbell agreed to serve as Grand Marshal of the Get Alarmed campaign. Fire safety programs, public service announcements, and printed fire prevention information will be available during the entire year. The programs will be designed to bring special seasonal messages.

WHO WILL BE INVOLVED IN THIS PROGRAM?

Citizens statewide will be involved in the Get Alarmed Program. Grand Marshal Iris Campbell appointed a coordinator in each county who will work with the fire service, community organizations, churches, schools and businesses in a partnership effort. The Division of State Fire Marshal of the Budget and Control Board will sponsor and support the campaign at the state level. The General Assembly approved funding for the program in the 1988/89 Appropriations Bill. With this funding, the Division of State Fire Marshal will provide 100 smoke detectors to each county. The county will be challenged to provide additional smoke detectors to meet the needs of all high risk citizens. Printed fire safety information and public service announcements will be furnished by the Division of State Fire Marshal for use in all counties.

HOW CAN YOU GET MORE INFORMATION ABOUT THE CAMPAIGN?

Information about the Get Alarmed campaign is available from your local fire department or the Division of State Fire Marshal, Suite 810, 1201 Main Street, Columbia, SC 29201; telephone: 737-0660.
(803)

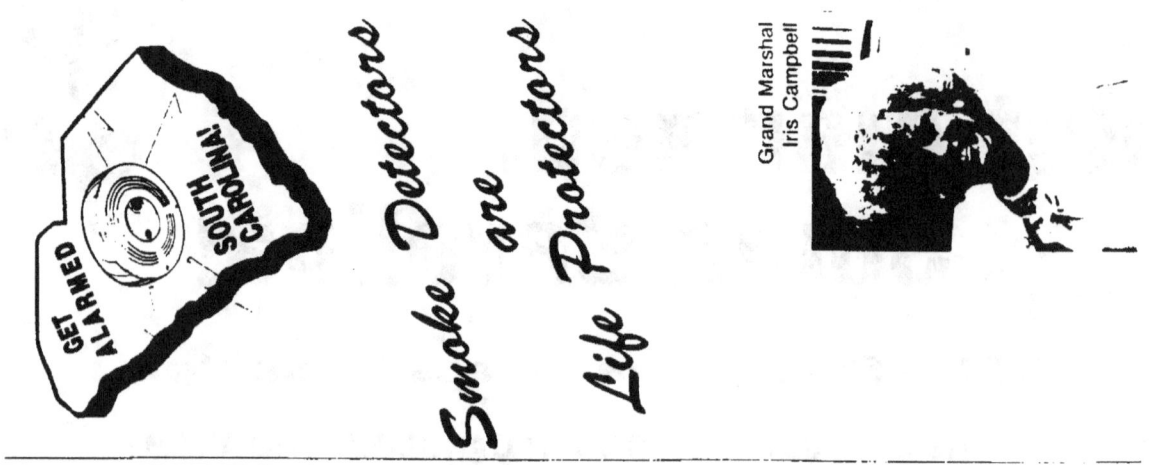

GET ALARMED SOUTH CAROLINA!

Smoke Detectors are Life Protectors

Grand Marshal
Iris Campbell

GET ALARMED SOUTH CAROLINA!

For more information, contact:
Division of State Fire Marshal
Budget and Control Board
Suite 810, 1201 Main Street
Columbia, SC 29201
(803-737-0660)
—or—
Your Local Fire Department

IF YOUR SMOKE DETECTOR DOESN'T WORK PROPERLY
THE SILENCE...COULD BE

TEST.
Test your smoke detector at least once a month. Push the test button or use smoke.

CLEAN.
Clean your detector at least once a year. Dust with a vacuum cleaner.

REPLACE.
Replace the battery each year. Use the battery type listed on the detector.

SAFETY TIPS:

HELP SAVE YOUR LIFE & PROPERTY FROM FIRE:

1 For minimum protection, install a smoke detector outside of each bedroom or sleeping area in your home and keep your bedroom doors closed while you are asleep.

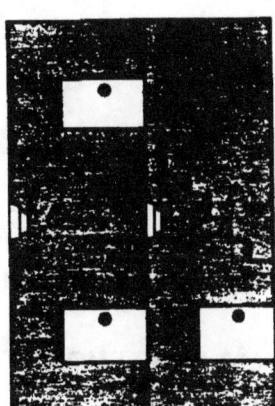

2 Keep your smoke detectors properly maintained. Test them at least once each month to insure that the detectors are working properly. Batteries in battery-operated detectors should be changed at least once yearly. Use only the type of batteries recommended on the detector.

3 If your smoke detector sounds an alarm when no smoke is present, consult with the manufacturer or with your local fire department. If smoke from cooking materials causes the detector to sound an alarm, do not remove the batteries or disconnect the power source. Simply fan the smoke away from the detector until the alarm stops. If this happens frequently, it may be necessary to relocate the detector or to install a different type of detector.

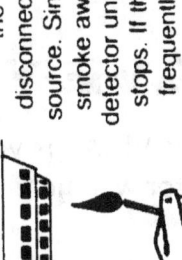

4 Develop an escape plan and review the plan with all members of the family frequently. Be aware that children and elderly people may need special assistance should fire occur. Establish a meeting place outside the house for all members of the family to insure that everyone got out safely. When fire occurs, get out of the house and use a neighbor's telephone to notify the fire department.

Contact your local fire department or State Fire Marshal for further information on fire prevention and fire safety.

LIVING IN A HOME WITHOUT SMOKE DETECTORS IS RISKY BUSINESS!

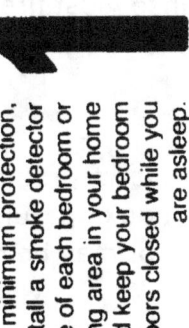

When fire occurs in your home, your chances for survival are two times better when smoke detectors are present than when they are not.

Smoke detectors, when properly installed and maintained (following manufacturer's directions), provide early warning when fires occur. Early warning increases your chances for survival and allows the fire department to save more of your property.

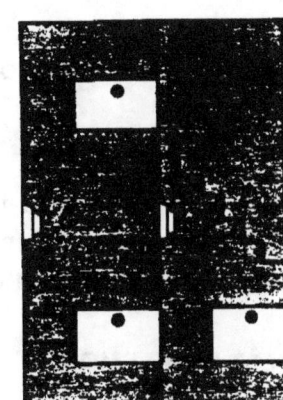

In support of smoke detector effectiveness, many cities and states have laws requiring the installation of smoke detectors in dwellings. Check with your local fire department or State Fire Marshal for further information.

Smoke Detectors
Save Lives ..

**GRAND MARSHAL
IRIS CAMPBELL**

THEY COULD SAVE YOURS!

In the event you have a house fire, SURVIVAL is your major concern. A properly installed and maintained smoke detector will increase your chances of survival by at least 2½ times.

Can you afford to take the chance of not having an early warning device? Protect your family and yourself. INSTALL AND MAINTAIN A SMOKE DETECTOR!

FOR MINIMAL PROTECTION

Install a smoke detector on each level of your home. Since most fire deaths occur when people are asleep, a smoke detector should be installed between the sleeping area and living area.

FOR MAXIMUM PROTECTION

Install a smoke detector in every room where a false alarm problem would not occur due to the presence of excessive steam or cooking fumes. In these problem areas (kitchens, bathrooms, etc.), a heat detector may be more suitable.

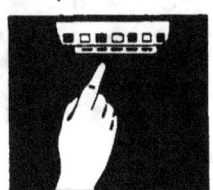

**TEST YOUR SMOKE
DETECTOR AT
LEAST ONCE A
MONTH**

**CLEAN TO
REMOVE DIRT AND
GREASE THAT COULD
INTERFERE WITH PROPER
OPERATION.**

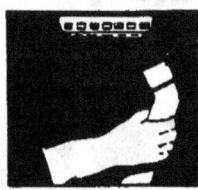

**BATTERIES NEED TO
BE REPLACED AT LEAST
ANNUALLY OR WHEN
DETECTOR BEGINS TO
"CHIRP".**

For more information, contact:
Division of State Fire Marshal — or — Your Local Fire Department
Budget and Control Board
Suite 810, 1201 Main Street
Columbia, SC 29201
(803) 737-0660

HOLIDAY SAFETY

GRAND MARSHAL
IRIS CAMPBELL

Holiday homes are beautiful. The season is joyous. And, there is ever present a threat of fire if safety practices are not observed.

CANDLES

Winter holidays and candles go together. They symbolize joy, warmth, security and light. If you use candles or other open flame in your household decorations, remember these safety tips. Keep flames well away from anything that will burn — upholstered furniture, curtains, lampshades, and flammable holiday decorations. Never leave a flame unattended, even for a moment, and keep children and pets away from all fires. If a candle is used in an arrangement with greens, be sure it's in a stable glass enclosure. Keep the greenery moist and discard as soon as it begins to dry.

CHRISTMAS TREES

Christmas trees are the most prevalent and the most beautiful of all decorations. Whether you have a cut tree or a living (container) tree, keep it well watered. No matter how charming the tradition, never use live candles on or near the tree. Christmas lights should be approved by Underwriters Laboratories as shown by the UL tag on the cord. Never use strings of lights that have frayed or broken cords. Do not overload electric outlets or overuse extension cords. Demanding too much current from your circuits can make wiring overheat and start fires inside your walls. Purchase only living trees with root balls or freshly cut trees. Some cut trees were harvested months earlier and are dangerously dry by the time you shop for your tree. Test freshness by closing your thumb and forefinger around a branch and pulling toward the branch tip. The needles should feel soft and springy, and very few should drop to the ground. When you get the tree home, cut about an inch off the bottom of the trunk and place the tree in water. The cutting step is important because a cut tree tries to "heal" its cut by sealing it with its own sap which prevents the tree from drinking. Use a stand that holds plenty of water so the tree will not dry out. A tree can become a flaming torch in a few seconds with a fire so hot that it can engulf an entire room in less than two minutes.

EXITS

Don't block doorways or hallways with your tree. In a fire, seconds count and every possible exit should be available.

FIRE SAFETY GIFTS

A novel gift that says you really care is a fire safety gift. How about a smoke detector, fire escape ladders, or portable fire extinguishers for your family and friends? If you need more information about these gift ideas, call your local fire department or the Division of State Fire Marshal, 737-0660. .

Don't let a fire end your wonderful holiday festivities.

Plan Your Escape

GRAND MARSHAL
IRIS CAMPBELL

Practice Exit Drills in the Home

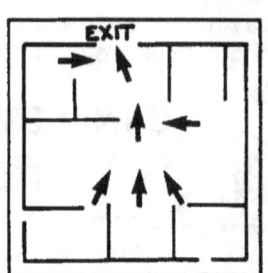

Make sure all family members know at least two ways out of every room. Draw a floor plan of your home and discuss the ways to exit each room with your family.

Designate a meeting place outside so you can determine if everyone has escaped. **DO NOT REENTER!**

Children should be told the importance of escaping and not hiding from the fire. Emphasize that they should get out fast and go directly to the meeting place.

Use a neighbor's telephone to call your fire department.

You should never, UNDER ANY CIRCUMSTANCES, let anyone reenter the burning building. Fire department personnel are trained and equipped to enter if necessary.

After your plan has been discussed with each family member, practice your exit drill regularly so that everyone will respond calmly and effectively in the event of a fire.

For more information contact:

Division of State Fire Marshal
Budget and Control Board
Suite 810, 1201 Main Street
Columbia, South Carolina 29201
(803) 737-0660
— or— Your Local Fire Department

Stay Warm Safely

**GRAND MARSHAL
IRIS CAMPBELL**

■ Never leave a kerosene heater unattended while in use. ■ Never operate a kerosene heater in a room without proper ventilation. ■ Never use gasoline or any other fuel in a heater designed for kerosene — this could cause an explosion. ■ Never store fuel inside your home — always refuel heater

outside, wiping up any spills. ■ Never place portable heaters close to combustibles such as furniture, walls, drapes, etc. Check owner manual for proper clearance instructions. ■ Never leave electric heaters plugged "in" when not in use. ■ Never build a fire in your fireplace without having a screen or glass door to protect combustibles from flying sparks. Inspect chimney for creosote build-up and clean if necessary. ■ Never empty ashes into paper or plastic containers. Discard them outside in metal containers.

For more information, contact:
**Division of State Fire Marshal
Budget and Control Board
Suite 810, 1201 Main Street
Columbia, SC 29201
(803) 737-0660
— or —
Your local fire department**

Their Life Is Your Responsibility

Young children and people over the age of 65 need to be especially aware of fire safety since they are statistically far more likely to die in home fires than the general population.

In many cases, they depend upon others for their safety. Make it YOUR RESPONSIBILITY to make sure they are as safe as possible.

- Develop an escape plan and make sure everyone knows at least two ways out of each room. When small children or the elderly are in your household, appoint another member of the family to assist them in escaping in the event of a fire. Handicapped citizens may require special assistance. Be aware of their needs.

- Never allow anyone to smoke in bed. Some senior citizens may need supervision while smoking. Always use large, deep ashtrays to avoid a lit cigarette from dropping on combustibles.

- Never leave the very young or very old unattended. It only takes a second for a mishap to occur.

- Never let children play with matches or lighters. THEY ARE NOT TOYS.

- Kitchen safety is a must when the young and elderly are present. Keep pot handles turned to the back of the stove so they cannot be grabbed or knocked off accidentally. Turn off the oven or burners as soon as you are finished cooking.

GET ALARMED, SOUTH CAROLINA

BROUGHT TO YOU BY YOUR

LOCAL FIRE DEPARTMENT

AND THE

SOUTH CAROLINA DIVISION OF STATE FIRE MARSHAL

SCHOOL'S OUT!
Prepare Your Children for a Fire Safe Summer

School's out and many children are left alone during the summer months while parents are working. Sad, but true, FIRE KILLS MORE CHILDREN THAN ANY OTHER HOME ACCIDENT.

Be sure that your child understands what to do if a fire occurs and you are not home. Discuss the following safety tips with your children to help ensure a safe and fun summer.

—If a fire breaks out, LEAVE THE HOUSE IMMEDIATELY — call the fire department from a neighbor's phone. NEVER Go Back Inside during a burning home for pets or belongings.

—If clothing catches fire, STOP — never run — this only stimulates the fire; DROP — to the ground and cover your face; and ROLL — over and over until the fire is out.

—NEVER PLAY WITH MATCHES OR LIGHTERS; they are not toys. Parents should keep these articles out of reach.

—Cooking should be left to older children and even then stress SAFETY while in the kitchen; NEVER leave the stove unattended while cooking. Teach young children to NEVER ATTEMPT TO COOK.

—CRAWL LOW IN SMOKE. In most cases, smoke is the KILLER. Teach your children to CRAWL to safety — UNDER THE SMOKE.

—Make an exit plan and practice — KNOW AT LEAST TWO WAYS OUT of every room. If a fire breaks out, everyone will know how to escape to safety.

"GET ALARMED, SOUTH CAROLINA!"
For More Information
Contact Your Local Fire Department
or the
South Carolina Division of State Fire Marshal

CHECK YOUR HOT SPOTS

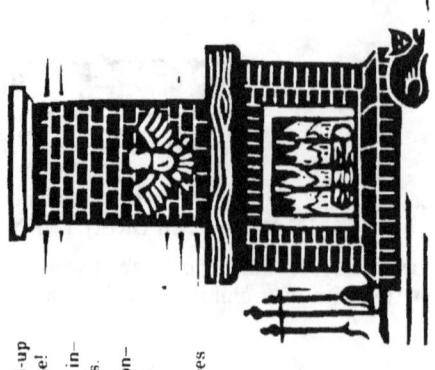

Electric Space Heaters

- Space heaters need their space! Keep combustibles at least three feet away from each heater.
- **Check Your Hot Spots!** When buying a heater, look for a thermostat control mechanism and a switch that automatically shuts off the power if the heater falls over.
- Heaters are not dryers or tables! Don't dry or store objects on top of your heater.
- A working smoke detector *doubles* your chances of surviving a fire.
- Practice a home escape plan frequently with your family.

Kerosene Heaters

- **Do not** fill your heater with gasoline or camp stove fuel - both explode easily.
- Buy crystal clear, K-1 kerosene only. There's no substitute for safety.
- **Check your Hot Spots!** Allow heater to cool before refueling and only refuel outdoors.
- Use the kerosene heater in a well ventilated room.
- Use only UL approved heaters — check with your local fire department on the legality of kerosene heater use in your community.

Fireplaces

- Keep fire where it belongs - in the fireplace! Make sure you have a screen large enough to catch flying sparks and rolling logs.
- Clean your chimney regularly - creosote build-up can ignite your chimney, roof, the whole house!
- **Check your hot spots!** Have your chimney inspected annually for damage and obstructions.
- Store cooled ashes in a tightly sealed metal container. Cardboard boxes and paper bags can quickly catch fire.
- A working smoke detector *doubles* your chances of surviving a fire.

Woodstoves

- Don't blow your stack over fire safety! **Carefully** follow manufacturer's installation and maintenance instructions.
- Use seasoned wood for fuel, not green wood, artificial logs or trash.
- **Check Your Hot Spots!** Clean pipes & chimneys annually and check monthly for damage or obstructions.
- Keep combustible objects at least three feet away from your woodstove.
- A working smoke detector *doubles* your chances of surviving a fire.

"GET ALARMED, SOUTH CAROLINA!"

For More Information

Contact Your Local Fire Department

or the

South Carolina Division of State Fire Marshal

"Put A Lid On Cooking Fires"

If grease ignites while you're cooking, you must smother the fire quickly and carefully. If the fire is in a pan on the stovetop, smother it by sliding a lid or large pan onto the top of the fire.

NEVER throw water or flour on a flaming pan. Water will spread the flames and flour will actually explode. Do not try to pick up the pan to carry it outside. Do not turn on the exhaust fan over the stove. If the fire is in the oven, turn off the heat and leave the door closed to cut off the fire's air supply.

If the fire is too large to safely cover with a lid or for your fire extinguisher, leave the house immediately and call the fire department from a neighbor's phone. Trying to fight a fire yourself when it's too large can cause far more extensive damage and result in serious injury.

Practice good fire prevention by keeping your cooking appliances clean. Grease build-up can flare up easily. Keep combustibles, paper towels, dish towels, paper bags, away from the stove or other heat source.

Do not wear long flowing sleeves while cooking. The clothing could easily touch a burner and ignite.

Always turn pot handles inward to avoid spills.

Keep children away from stoves.

It's always a good idea to check stoves and other appliances before going to bed or leaving your home to make sure that everything has been turned off.

"GET ALARMED, SOUTH CAROLINA!"

For More Information

CONTACT YOUR LOCAL FIRE DEPARTMENT

or the

SOUTH CAROLINA DIVISION OF STATE FIRE MARSHAL

Don't Let A Fire Take The Bang Out Of Summer

Along with warm weather, summer brings fun with cookouts, fireworks, and other outside activities. Keep your summer safe by observing these tips.

While using fireworks,

— Always Read and Follow Directions

— Have An Adult Present

— Do Not Use Near Dry Grass or Other Flammable Materials

— Never Point or Throw Fireworks at Another Person — Fireworks Are Not Toys

— Never Attempt To Relight or "Fix" Fireworks

— Keep a Safe Distance — Fireworks Can Hurt

Precautions should also be taken to ensure that your cookout doesn't turn into a BURN-OUT!

— Keep Grill Away From Combustible Walls or Materials

— Always Use Long-Handled Utensils

— Keep Water Nearby to Douse Small Fires

— Keep Children and Pets Away From Hot Grill

GET ALARMED, SOUTH CAROLINA! Brought To You By Your

LOCAL FIRE DEPARTMENT and the SOUTH CAROLINA DIVISION OF STATE FIRE MARSHAL

FREDDIE

THE FIRE-LESS FELINE

FREDDIE, THE FIRE-LESS FELINE

One night when it was very dark and very quiet, the Jones family was sound asleep in their pretty little house in South Carolina. Mom and Dad, Jamie, and Suzie, were sleeping in their upstairs bedrooms. All of a sudden, the smoke detector in the parent's bedroom made a loud noise, telling the family that it smelled smoke. Mom called out to Suzie and Jamie just as they were all awakened suddenly by the loud beeping of the smoke alarm in each of the bedrooms. Many times the family had practiced what they would do if they had a real fire. Each went to his bedroom door, felt it, and when they found that it was not hot, they opened it very carefully. They could smell smoke, and they could see it up near the ceiling, down near the stairs, and it was slowly coming down the hall, up high but creeping in around the top of the doors. They knew the closest way out would be the front door, so they closed their bedroom doors and got down on their hands and knees because they remembered to crawl low in smoke. When they were safely outside, they all went to the mail box, making sure that all four family members were there. When they were all counted, Dad went next door and called the fire department.

Jamie started to cry. His new little kitten, Freddie, was sleeping in a basket in the kitchen. Jamie knew he could not go back inside to save Freddie. His parents and his teachers had told him many times that if you had a fire in your home, you must go to the family meeting place (the mail box) and stay there. And never, never go back inside. But, Jamie felt so sad, and he knew that Freddie was sooo scared.

Soon the big fire trucks came. The sirens were making loud noises. The firemen were very busy with big hoses. They had big ladders. Jamie and Suzie were scared. Inside, Freddie woke up. He smelled smoke. Even though the smoke was still high in the room, it made his eyes and throat hurt. He could see the smoke all around. But Freddie was so little, and he did not know that he should not hide in a fire because the firemen might not find him, so he scurried under a door and ran into a closet and meowed and meowed--he was so scared. Where was Jamie? Where was Suzie? Why didn't they come and get him? Freddie went way back into a corner and cried and cried. All of a sudden, the closet door opened and Freddie saw the biggest pair of boots he had ever seen. When he looked up, there was a man so big that Freddie couldn't see his head. The man bent over and picked Freddie up in his big gloves. Then Freddie saw that the man had on a funny hat, like a helmet, and a mask on his face, like a diver. Freddie was no longer afraid because he knew the man was a fireman. Jamie had read stories about firemen, and he had showed pictures to Freddie.

The smoke was getting bad and the firemen had put a lot of water on the fire. Freddie closed his eyes and ducked his head into the fireman's big gloves. The fireman took Freddie outside and gently gave him to Jamie. Jamie stopped crying. He was so happy to know that his pet was safe. He thanked the fireman over and over. He was glad the fireman found him even though he had gone into the closet.

The Jones family learned that the fire had started in the laundry room. Because the smoke

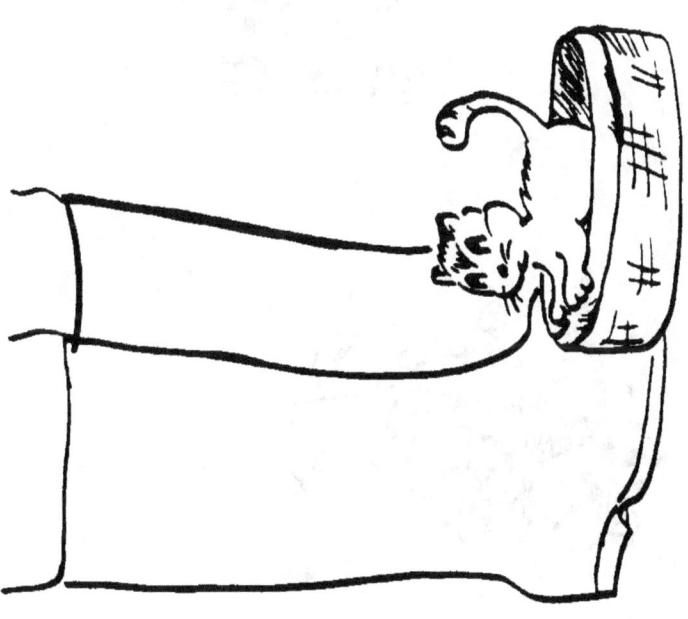

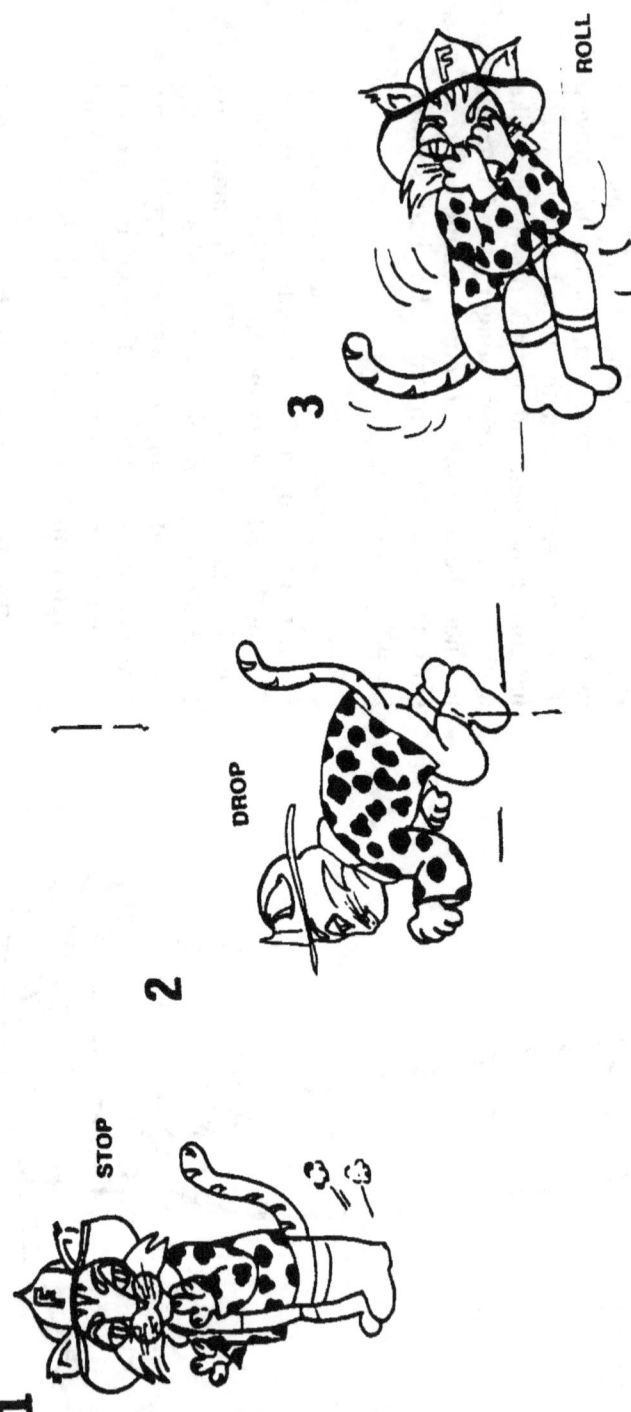

EVERY HOME NEEDS A SMOKE DETECTOR

detector had smelled the smoke and given an early warning, the firefighters were able to put the fire out and most of the family's furniture and other belongings were saved. Everyone in the family was happy because their home was not lost. Freddie, from that day on, never forgot the fireman who had saved his life. Freddie decided that when he grew up he would be a fire cat and work in a fire department.

So, when Freddie grew older, he decided the time had come for him to become a fire prevention cat. He knew that dogs live in fire departments--Dalmatian Dogs, white with black spots on them. How could Freddie convince the firemen that a cat could live in a fire station and teach children about fire safety? Would the firemen want a cat in the fire house? If a cat knew ALL about fire safety--more than a dog, then, maybe the firemen would let him live with them and teach boys and girls.

So, Freddie learned all that he could about fire safety. He learned what to do if your clothes catch on fire: Stop, Drop, and Roll! He learned that children should never play with matches or cigarette lighters. He learned about smoke detectors and how to change their batteries. He learned that children should never play around stoves. He learned about fire drills and exit plans.

Maybe if he looked just a little more like a Dalmatian Dog, the firemen would accept him. So, he went to the store and bought some cloth that looked like fur--white with black spots on it. Mrs. Jones made him a nice jacket. And, he got some yellow boots. He loved big boots like the ones the fireman wore when he saved his life. He got a nice red fireman's hat. When he looked in the mirror, he was really happy, and he knew the time had come for him to go to the fire department. He knew that little children needed to learn about fire safety. He had heard Mr. and Mrs. Jones talk about many children dying in fires because they did not know how to be safe. Freddie wanted to teach them, so he told Jamie and Suzie good-bye and he set out to find a fire department.

Pretty soon, he saw the fire station. How excited he became when he saw the bright shiny fire trucks! One of the trucks was outside and the firemen were washing it. Freddie walked right up and asked, "Where can I find the fire chief? My name is Freddie, the Fire-less Feline." Everyone knows that a feline is a cat. Freddie thought that "feline" sounded better than just plain "cat". The firemen laughed and looked surprised when they saw a cat in a spotted jacket, with a fireman's hat and yellow boots.

About that time, around the corner came a big black and white spotted dog barking. "Hold on there, dog," said Freddie, "I have come to see the fire chief and to teach children about fire safety. You and I must become friends." The dog looked puzzled, but he let Freddie pass. Soon, a nice looking man in a red car drove up to the station, and one of the firemen said, "Here's the chief now." The chief invited Freddie into his office. Freddie's heart was pounding, and he was just a little bit afraid. But he told the fire chief about his experience when he was a little kitten and how the fireman had saved his life. He told the fire chief he had worked very hard to learn all about fire safety and that he wanted to help children learn to be safe. The fire chief was impressed that a cat could know so much--even more than a dog. He agreed to let Freddie live at the fire station. Freddie's job would be to talk to children when they came to visit the fire house. Sometimes Freddie would go to schools and day care centers to talk about fire prevention. Sometimes he would ride in the big fire truck in parades. And, sometimes he would go to fire prevention fairs. One day he even went to a tall building to see the State Fire Marshal. Freddie was so happy.

And that's how a little gray kitten became Fireless Freddie, a fire safety hero who teaches children in South Carolina!

Division of State Fire Marshal
1201 Main Street, Suite 810
Columbia, SC 29201
Telephone: 803/737-0660

State of South Carolina
Division of State Fire Marshal
Budget and Control Board

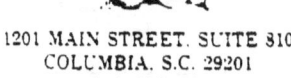

CARROLL A. CAMPBELL, JR., CHAIRMAN
GOVERNOR
GRADY L. PATTERSON, JR.
STATE TREASURER
EARLE E MORRIS, JR.
COMPTROLLER GENERAL

1201 MAIN STREET, SUITE 810
COLUMBIA, S.C. 29201
(803) 737-0660

JAMES M. WADDELL, JR.
SENATE FINANCE COMMITTEE
ROBERT N. McLELLAN, CHAIRMAN
WAYS AND MEANS COMMITTEE
JESSE A. COLES, JR., Ph.D.
EXECUTIVE DIRECTOR

RICHARD S. CAMPBELL, P.E.
STATE FIRE MARSHAL

Mary Lee Maiden
Public Information Manager, 737-0660

RADIO PSA :30 SCRIPT
SUMMER FUN
Summer fun and summer chores can cause serious accidents.

Whenever using any engine fueled by gasoline--whether a boat, minibike
or lawnmower--use extreme caution.

The Division of State Fire Marshal wants you to know that gasoline
creates invisible vapors that can drift away from the source and
explode without warning.

So never smoke while using gasoline.

Turn off engines and let them cool before fueling. Then move the
engine at least 10 feet from where you fueled it before restarting.

Never use gasoline for cleaning or to start a fire.

So start your engines--but carefully.

This message from the "Get Alarmed, South Carolina! fire safety campaign.

#########

Please play during July and August.

State of South Carolina
Division of State Fire Marshal
Budget and Control Board

CARROLL A. CAMPBELL, JR., CHAIRMAN
GOVERNOR
GRADY L. PATTERSON, JR.
STATE TREASURER
EARLE E. MORRIS, JR.
COMPTROLLER GENERAL

1201 MAIN STREET, SUITE 810
COLUMBIA, S.C. 29201
(803) 737-0660

JAMES M. WADDELL, JR.
SENATE FINANCE COMMITTEE
ROBERT N. McLELLAN, CHAIRMAN
WAYS AND MEANS COMMITTEE
JESSE A. COLES, JR., Ph.D
EXECUTIVE DIRECTOR

RICHARD S. CAMPBELL, P.E.
STATE FIRE MARSHAL

Contact: Mary Lee Maiden
Public Information Manager, 737-0660

RADIO PSA :30 SCRIPT:
SMOKE DETECTORS

Most fatal fires happen at night when you and your family are

asleep--when seconds can make the difference between life and death.

Are you protected?

If you have one or more working smoke detectors in your home, you

double your chances of surviving a fire! That's protection!

And smoke detectors are inexpensive and easy to install.

Smoke detectors--they're life protectors.

This message from the "Get Alarmed, South Carolina! fire safety campaign.

#############

State of South Carolina
Division of State Fire Marshal
Budget and Control Board

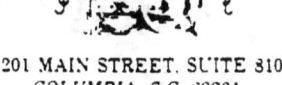

CARROLL A. CAMPBELL, JR., CHAIRMAN
GOVERNOR
GRADY L. PATTERSON, JR.
STATE TREASURER
EARLE E. MORRIS, JR.
COMPTROLLER GENERAL

1201 MAIN STREET, SUITE 810
COLUMBIA, S.C. 29201
(803) 737-0660

JAMES M. WADDELL, JR.
SENATE FINANCE COMMITTEE
ROBERT N. McLELLAN, CHAIRMAN
WAYS AND MEANS COMMITTEE
JESSE A. COLES, JR., Ph.D.
EXECUTIVE DIRECTOR

RICHARD S. CAMPBELL, P.E.
STATE FIRE MARSHAL

Mary Lee Maiden, 737-0660
Public Information Manager

RADIO PSA :45 SCRIPT

SMOKE DETECTORS MESSAGE FOR PARENTS

The Division of State Fire Marshal has a special message for parents.

Most fatal fires occur at night, when you and your family are

asleep-- and when seconds can mean the difference between life and

death.

Smoke detectors, when properly installed and maintained, double your

chances of surviving a fire.

Smoke detectors are inexpensive. Put at least one on every level of

your home.

Remember to test your smoke detectors monthly and to change the

battery at least once a year--on your birthday.

        ....because you don't need to leave your family's fire

safety to chance.

Smoke detectors--they're life protectors.

This message from the "Get Alarmed, South Carolina! fire safety campaign.

##########

State of South Carolina
Division of State Fire Marshal
Budget and Control Board

CARROLL A. CAMPBELL, JR., CHAIRMAN
GOVERNOR

GRADY L. PATTERSON, JR.
STATE TREASURER

EARLE E. MORRIS, JR.
COMPTROLLER GENERAL

1201 MAIN STREET, SUITE 810
COLUMBIA, S.C. 29201
(803) 737-0660

JAMES M. WADDELL, JR.
SENATE FINANCE COMMITTEE

ROBERT N. McLELLAN, CHAIRMAN
WAYS AND MEANS COMMITTEE

JESSE A. COLES, JR., Ph.D
EXECUTIVE DIRECTOR

RICHARD S. CAMPBELL, P.E.
STATE FIRE MARSHAL

Mary Lee Maiden
Public Information Manager, 737-0660

RADIO PSA :15 SCRIPT:
SMOKE DETECTOR INSTALLATION

The Division of State Fire Marshal's "Get Alarmed, South Carolina!

campaign reminds you that...most fatal fires take place at night when

you and your family are asleep...and that smoke detectors double your

chances of surviving a fire.

Smoke detectors are both inexpensive and easy to install.

Smoke detectors--they're life protectors.

This message from the "Get Alarmed, South Carolina! fire safety campaign.
#########

(Please play during July and August)

State of South Carolina
Division of State Fire Marshal
Budget and Control Board

CARROLL A. CAMPBELL, JR., CHAIRMAN
GOVERNOR
GRADY L. PATTERSON, JR.
STATE TREASURER
EARLE E. MORRIS, JR.
COMPTROLLER GENERAL

1201 MAIN STREET, SUITE 810
COLUMBIA, S.C. 29201
(803) 737-0660

JAMES M. WADDELL, JR.
SENATE FINANCE COMMITTEE
ROBERT N. McLELLAN, CHAIRMAN
WAYS AND MEANS COMMITTEE
JESSE A. COLES, JR., Ph.D
EXECUTIVE DIRECTOR

RICHARD S. CAMPBELL, P.E.
STATE FIRE MARSHAL

Mary Lee Maiden
Public Information Manager, 737-0660

RADIO PSA :60 SCRIPT
SMOKE DETECTOR MAINTENANCE

The experts at the U. S. Fire Administration tell us that smoke
detectors double your chance of surviving a house fire. But
installing the detector is not enough. You have to keep it working
properly. If it sounds an alarm when there's no smoke, consult your
local fire department. Check and clean the detector monthly, and
change the batteries every year. Change the detector's location if it
always reacts to cooking smoke. And develop a fire plan with your
family. Practice your escape route from time to time. Make sure
everyone is prepared for the worst disaster. That way, you'll be
better able to survive if it does arrive.

This message from the "Get Alarmed, South Carolina! fire safety campaign.

#########

State of South Carolina
Division of State Fire Marshal
Budget and Control Board

CARROLL A. CAMPBELL, JR., CHAIRMAN
GOVERNOR
GRADY L. PATTERSON, JR.
STATE TREASURER
EARLE E. MORRIS, JR.
COMPTROLLER GENERAL

1201 MAIN STREET, SUITE 810
COLUMBIA, S.C. 29201
(803) 737-0660

JAMES M. WADDELL, JR.
SENATE FINANCE COMMITTEE
ROBERT N. McLELLAN, CHAIRMAN
WAYS AND MEANS COMMITTEE
JESSE A. COLES, JR., Ph.D
EXECUTIVE DIRECTOR

RICHARD S. CAMPBELL, P.E.
STATE FIRE MARSHAL

Mary Lee Maiden
Public Information Manager, 737-0660

RADIO PSA: :30 SCRIPT:
GET ALARMED CAMPAIGN

Governor Campbell proclaimed 1988 as "Fire Safety Awareness Year in
South Carolina. "

First Lady Iris Campbell serves as Grand Marshal of "Get Alarmed,
South Carolina!", a special fire safety awareness campaign.

The campaign is sponsored by the Division of State Fire Marshal of the
Budget and Control Board.

Stay tuned to this station where you will hear special fire safety
messages that could save your life. Contact your local fire
department for more information.

########

(Use first and throughout remainder of 1988 as a forerunner for other
messages.)

State of South Carolina
Division of State Fire Marshal
Budget and Control Board

CARROLL A. CAMPBELL, JR., CHAIRMAN
GOVERNOR
GRADY L. PATTERSON, JR.
STATE TREASURER
EARLE E. MORRIS, JR.
COMPTROLLER GENERAL

1201 MAIN STREET, SUITE 810
COLUMBIA, S.C. 29201
(803) 737-0660

JAMES M. WADDELL, JR.
SENATE FINANCE COMMITTEE
ROBERT N. McLELLAN, CHAIRMAN
WAYS AND MEANS COMMITTEE
JESSE A. COLES, JR., Ph.D
EXECUTIVE DIRECTOR

RICHARD S. CAMPBELL, P.E.
STATE FIRE MARSHAL

Mary Lee Maiden
Public Information Manager, 737-0660

RADIO PSA :45 SCRIPT
STAYING ALIVE

Improving your chances of surviving a fire in your home can be easy.
Installing a smoke detector and keeping it properly maintained is the
smart way.

First and foremost, follow package instructions when installing the
device. Place the detectors in appropriate locations, near bedrooms
and living areas.

Even if the alarm sounds because of cooking smoke, resist the
temptation to remove the battery. You need the detector working at all
times.

The State Fire Marshal recommends buying detectors for family and
friends who are without them. It's one gift that should get a warm
welcome.

This message from the "Get Alarmed, South Carolina! fire safety campaign.

##########

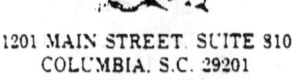

State of South Carolina

Division of State Fire Marshal

Budget and Control Board

CARROLL A. CAMPBELL, JR., CHAIRMAN
GOVERNOR
GRADY L. PATTERSON, JR.
STATE TREASURER
EARLE E. MORRIS, JR.
COMPTROLLER GENERAL

1201 MAIN STREET, SUITE 810
COLUMBIA, S.C. 29201
(803) 737-0660

JAMES M. WADDELL, JR.
SENATE FINANCE COMMITTEE
ROBERT N. McLELLAN, CHAIRMAN
WAYS AND MEANS COMMITTEE
JESSE A. COLES, JR., Ph.D
EXECUTIVE DIRECTOR

RICHARD S. CAMPBELL, P.E.
STATE FIRE MARSHAL

Mary Lee Maiden
Public Information Manager, 737-0660

RADIO PSA :30 SCRIPT
Survival

According to the U. S. Fire Adminisration, smoke detectors decrease
your risk of dying in a home fire by as much as fifty percent.

The earlier the fire is discovered, the better the chance for
survival. Smoke detectors should be on every level of your home.

Test your detector at least once a month and change the battery every
year. A chirping noise means the detector needs a new battery.

It's a life protector, a smoke detector.

A message from this station and your local fire department.

This message from the "Get Alarmed, South Carolina! fire safety campaign.
#########

State of South Carolina

Division of State Fire Marshal
Budget and Control Board

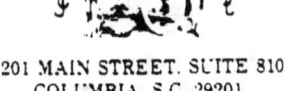

CARROLL A. CAMPBELL. JR., CHAIRMAN
GOVERNOR
GRADY L. PATTERSON. JR.
STATE TREASURER
EARLE E. MORRIS. JR.
COMPTROLLER GENERAL

1201 MAIN STREET. SUITE 810
COLUMBIA. S.C. 29201
(803) 737-0660

JAMES M. WADDELL. JR.
SENATE FINANCE COMMITTEE
ROBERT N. McLELLAN. CHAIRMAN
WAYS AND MEANS COMMITTEE
JESSE A. COLES. JR.. Ph.D
EXECUTIVE DIRECTOR

RICHARD S. CAMPBELL. P.E.
STATE FIRE MARSHAL

Mary Lee Maiden
Public Information Manager, 737-0660

RADIO PSA :30
AFTER SCHOOL KIDS

In this country, millions of children care for themselves after school
while their parents work. All children should learn about fire
safety, but these "latchkey" children especially need to be taught how
to protect themselves from fire.

The State Fire Marshal wants to remind parents to teach their children
to leave matches and cigarette lighters alone.

If clothes catch on fire, stop, drop, and roll.

If burned by something hot, cool the burn with cool water, ever
anything greasy.

If a fire starts, leave the house immediately and call the fire
department from a neighbor's house.

This message from the "Get Alarmed, South Carolina! fire safety campaign.

##########

State of South Carolina
Division of State Fire Marshal
Budget and Control Board

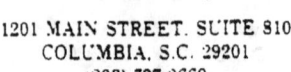

CARROLL A. CAMPBELL, JR. CHAIRMAN
GOVERNOR
GRADY L. PATTERSON, JR.
STATE TREASURER
EARLE E. MORRIS, JR.
COMPTROLLER GENERAL

1201 MAIN STREET, SUITE 810
COLUMBIA, S.C. 29201
(803) 737-0660

JAMES M. WADDELL, JR.
SENATE FINANCE COMMITTEE
ROBERT N. McLELLAN, CHAIRMAN
WAYS AND MEANS COMMITTEE
JESSE A. COLES, JR., Ph.D
EXECUTIVE DIRECTOR

RICHARD S. CAMPBELL, P.E.
STATE FIRE MARSHAL

Mary Lee Maiden
Public Information Manager, 737-0660

RADIO PSA :30 SCRIPT
FIRE AND DRINKING

Drinking and driving are a deadly combination. Recent publicity may
improve the terrible statistics.

But even when drinking at home, the State Fire Marshal reminds you
that alcohol puts you at risk for a deadly fire.

Studies have shown that more than one-third of people who die in fires
were legally drunk.

Drinking makes people careless, puts them to sleep and keeps them from
waking up to save themselves if a fire breaks out.

So stay sober and alert.

This message from the "Get Alarmed, South Carolina! fire safety campaign.

##########

State of South Carolina
Division of State Fire Marshal
Budget and Control Board

CARROLL A. CAMPBELL, JR., CHAIRMAN
GOVERNOR
GRADY L. PATTERSON, JR.
STATE TREASURER
EARLE E. MORRIS, JR.
COMPTROLLER GENERAL

1201 MAIN STREET, SUITE 810
COLUMBIA, S.C. 29201
(803) 737-0660

JAMES M. WADDELL, JR.
SENATE FINANCE COMMITTEE
ROBERT N. McLELLAN, CHAIRMAN
WAYS AND MEANS COMMITTEE
JESSE A. COLES, JR., Ph.D
EXECUTIVE DIRECTOR

RICHARD S. CAMPBELL, P.E.
STATE FIRE MARSHAL

Mary Lee Maiden
Public Information Manager, 737-0660

RADIO PSA :30 SCRIPT
CIGARETTE SMOKING FIRES

A very high incidence of fire deaths, injuries and property losses in this country is attributed to cigarette smoking. In South Carolina, the figures are increasing. The State Fire Marshal recommends these precautions.

Provide large, deep ashtrays for smokers.

Be sure ashes are cool before disposal or flush down toilet.

Most cigarette fires start when a hot cigarette drops onto upholstery, bedding, carpet, or clothing.

Never smoke in bed or anywhere that may tempt you to sleep.

Install extra smoke detectors in smokers' bedrooms.

Remember, if you smoke while intoxicated, you may never wake up.

If you must smoke, be careful.

This message from the "Get Alarmed, South Carolina! fire safety campaign.

##########

State of South Carolina
Division of State Fire Marshal
Budget and Control Board

CARROLL A. CAMPBELL, JR., CHAIRMAN
GOVERNOR
GRADY L. PATTERSON, JR.
STATE TREASURER
EARLE E. MORRIS, JR.
COMPTROLLER GENERAL

1201 MAIN STREET, SUITE 810
COLUMBIA, S.C. 29201
(803) 737-0660

JAMES M. WADDELL, JR.
SENATE FINANCE COMMITTEE
ROBERT N. McLELLAN, CHAIRMAN
WAYS AND MEANS COMMITTEE
JESSE A. COLES, JR., Ph.D
EXECUTIVE DIRECTOR

RICHARD S. CAMPBELL, P.E.
STATE FIRE MARSHAL

Mary Lee Maiden
Public Information Manager, 737-0660

RADIO PSA :30 SCRIPT
HOLIDAY SAFETY

The State Fire Marshal want you to keep your holidays merry and

bright.

Keep candles well away from evergreen cuttings, curtains, furniture or

anything else that might burn.

Keep children away from flames.

Be sure your tree is fresh and keep it fresh by adding water

regularly.

Don't block exits with your tree.

Don't overload circuits or overuse extension cords.

Use only UL approved Christmas lights.

Give smoke detectors to family and friends.

Have a safe holiday season.

 This message from the "Get Alarmed, South Carolina! fire safety campaign.

#########

(NEED GOOD COVERAGE DECEMBER)

State of South Carolina
Division of State Fire Marshal
Budget and Control Board

CARROLL A. CAMPBELL, JR., CHAIRMAN
GOVERNOR
GRADY L. PATTERSON, JR.
STATE TREASURER
EARLE E. MORRIS, JR.
COMPTROLLER GENERAL

1201 MAIN STREET, SUITE 810
COLUMBIA, S.C. 29201
(803) 737-0660

JAMES M. WADDELL, JR.
SENATE FINANCE COMMITTEE
ROBERT N. McLELLAN, CHAIRMAN
WAYS AND MEANS COMMITTEE
JESSE A. COLES, JR., Ph.D
EXECUTIVE DIRECTOR

RICHARD S. CAMPBELL, P.E.
STATE FIRE MARSHAL

Mary Lee Maiden
Public Information Manager, 737-0660

RADIO PSA :30 SCRIPT
FIRE DRILLS
Practice is important to anything that should be done well. Your kids

have fire drills at school. You may even have them at work.

But, has your family ever practiced a fire drill at home?

The State Fire Marshal wants you to get out safely in case of fire.

In a fire you are likely to be groggy from sleep and from the smoky

gases in the air.

Normal exits may be blocked by fire and smoke, so you need to know

another way out.

Would your family become frightened, confused, and do the wrong thing?

Or would they be able to escape safely?

Every home should have a complete fire escape plan, and the whole

family should practice it together.

This message from the "Get Alarmed, South Carolina! fire safety campaign.

#########

State of South Carolina
Division of State Fire Marshal
Budget and Control Board

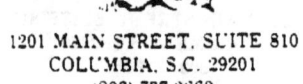

CARROLL A. CAMPBELL, JR., CHAIRMAN
GOVERNOR
GRADY L. PATTERSON, JR.
STATE TREASURER
EARLE E. MORRIS, JR.
COMPTROLLER GENERAL

1201 MAIN STREET, SUITE 810
COLUMBIA, S.C. 29201
(803) 737-0660

JAMES M. WADDELL, JR.
SENATE FINANCE COMMITTEE
ROBERT N. McLELLAN, CHAIRMAN
WAYS AND MEANS COMMITTEE
JESSE A. COLES, JR., Ph.D
EXECUTIVE DIRECTOR

RICHARD S. CAMPBELL, P.E.
STATE FIRE MARSHAL

Mary Lee Maiden
Public Information Manager, 737-0660

RADIO PSA :30 SCRIPT
STOP, DROP, AND ROLL

The State Fire Marshal wants you to know three simple words that can save your life--stop, drop, and roll. Even if you're always careful, some day your clothing might catch fire.

When clothing burns there seems to be a normal reaction to run--to run away from the fire or to run toward water or outside. But don't do it.

Stop right away, right where you are.

Drop immediately to the ground.

Roll back and forth to smother the fire against the ground or floor.

These actions will stop the fire as fast as possible, prevent the fire from getting worse and help control the severity of the burns.

Do it right. Stop, drop, and roll.

This message from the "Get Alarmed, South Carolina! fire safety campaign.

##########

State of South Carolina
Division of State Fire Marshal
Budget and Control Board

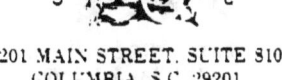

CARROLL A. CAMPBELL, JR., CHAIRMAN
GOVERNOR
GRADY L. PATTERSON, JR.
STATE TREASURER
EARLE E. MORRIS, JR.
COMPTROLLER GENERAL

1201 MAIN STREET, SUITE 810
COLUMBIA, S.C. 29201
(803) 737-0660

JAMES M. WADDELL, JR.
SENATE FINANCE COMMITTEE
ROBERT N. McLELLAN, CHAIRMAN
WAYS AND MEANS COMMITTEE
JESSE A. COLES, JR., Ph.D
EXECUTIVE DIRECTOR

RICHARD S. CAMPBELL, P.E.
STATE FIRE MARSHAL

Mary Lee Maiden
Public Information Manager, 737-0660

RADIO PSA :30 SCRIPT
HANDLE WITH CARE

All fires are dangerous, but electrical fires add the threat of electrocution. The State Fire Marshal wants you to respond safely to an electrical fire.

If an appliance catches fire, unplug it or shut the power off at the fuse or circuit breaker.

Do not use water on an electrical fire. Call the fire department.

Electrical fires can start inside walls, too. If you smell smoke or the odor of elecrical burning, call the fire department and get out of the house without delay.

Prevent electrical fires by not overloading electrical circuits. Even plugging in an extra fan can be too much. Don't take chances.

This message from the "Get Alarmed, South Carolina! fire safety campaign.

##########

State of South Carolina

Division of State Fire Marshal

Budget and Control Board

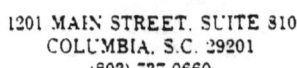

CARROLL A. CAMPBELL. JR. CHAIRMAN
GOVERNOR
GRADY L. PATTERSON. JR.
STATE TREASURER
EARLE E. MORRIS. JR.
COMPTROLLER GENERAL

1201 MAIN STREET. SUITE 810
COLUMBIA. S.C. 29201
(803) 737-0660

JAMES M. WADDELL. JR.
SENATE FINANCE COMMITTEE
ROBERT N. McLELLAN CHAIRMAN
WAYS AND MEANS COMMITTEE
JESSE A. COLES. JR. Ph.D
EXECUTIVE DIRECTOR

RICHARD S. CAMPBELL. P.E.
STATE FIRE MARSHAL

Mary Lee Maiden
Public Information Manager, 737-0660

```
RADIO PSA :30 SCRIPT
SMOKE KILLS
```

Most people who die in fires never see or feel the flames--they die
from breathing deadly smoke.

The State Fire Marshal wants you to stay low and go if you are caught
in a fire.

Smoke contains carbon monoxide that can dull your ability to think,
make you act irrationally and can kill. It also contains fumes from
the many plastic and artificial products used today.

Because smoke tends to hover near the top of a room, crawling is the
safest way to escape a fire. And don't worry about feeling silly.
It's the only way to get safely outside.

This message from the "Get Alarmed, South Carolina! fire safety campaign.

##########

State of South Carolina
Division of State Fire Marshal
Budget and Control Board

CARROLL A. CAMPBELL, JR., CHAIRMAN
GOVERNOR
GRADY L. PATTERSON, JR.
STATE TREASURER
EARLE E. MORRIS, JR.
COMPTROLLER GENERAL

1201 MAIN STREET, SUITE 810
COLUMBIA, S.C. 29201
(803) 737-0660

JAMES M. WADDELL, JR.
SENATE FINANCE COMMITTEE
ROBERT N. McLELLAN, CHAIRMAN
WAYS AND MEANS COMMITTEE
JESSE A. COLES, JR., Ph.D
EXECUTIVE DIRECTOR

RICHARD S. CAMPBELL, P.E.
STATE FIRE MARSHAL

Mary Lee Maiden
Public Information Manager, 737-0660

RADIO PSA :30 SCRIPT
COOKING FIRES

Thousands of people are seriously injured every year by cooking fires.

The State Fire Marshal suggests you keep your kitchen fire-safe and

know what to do if a fire starts.

Never leave cooking unattended.

If a fire starts in a pan on the stovetop, slide the pan's lid or a

larger pan on top to smother the fire.

If the fire's in the oven, turn off the heat and keep the oven door

closed.

You can use a fire extinguisher if you know how to use it. But, if

the fire's too big, don't try to fight it.

Leave the house right away and call the fire department from a

neighbor's phone.

 This message from the "Get Alarmed, South Carolina! fire safety campaign.

##########

State of South Carolina

Division of State Fire Marshal

Budget and Control Board

CARROLL A. CAMPBELL, JR., CHAIRMAN
GOVERNOR
GRADY L. PATTERSON, JR.
STATE TREASURER
EARLE E. MORRIS, JR.
COMPTROLLER GENERAL

1201 MAIN STREET, SUITE 810
COLUMBIA, S.C. 29201
(803) 737-0660

JAMES M. WADDELL, JR.
SENATE FINANCE COMMITTEE
ROBERT N. McLELLAN, CHAIRMAN
WAYS AND MEANS COMMITTEE
JESSE A. COLES, JR., Ph.D
EXECUTIVE DIRECTOR

RICHARD S. CAMPBELL, P.E.
STATE FIRE MARSHAL

Mary Lee Maiden
Public Information Manager, 737-0660

RADIO PSA :30 SCRIPT
STEAKS ARE FOR COOKING
It's easy to make steak taste great, especially when it's been grilled

outdoors. It's just as important to cook that steak carefully.

The Division of State Fire Marshal reminds you to add fire safety to

your barbecue recipe. Barbecue risk can be avoided with a little

common sense.

Wear clothing that won't flop against the hot grill.

Always use long-handled utensils intended for barbecuing.

The safest way to start your fire is with a bunch of crumpled

newspapers. It's low-tech but it works and it won't explode in your

face.

Keep a garden hose or large bucket of water nearby, just in case.

Have a safe, happy and delicious cookout.

This message from the "Get Alarmed, South Carolina! fire safety campaign.

##########

(Please play during July and August.)

www.ingramcontent.com/pod-product-compliance
Lightning Source LLC
Chambersburg PA
CBHW081223170526
45165CB00009B/2920